First published in paperback February 2014

Printed in the United States of America

First Edition

ISBN 0-615-96788-4

Morning Tea Press
LLC

Morning Tea Press, LLC
P.O. Box 282, Mount Vernon, VA 22121
www.morningteapress.com

UNIFIED FACILITIES CRITERIA (UFC)

HIGH PERFORMANCE AND SUSTAINABLE BUILDING REQUIREMENTS

U.S. ARMY CORPS OF ENGINEERS

NAVAL FACILITIES ENGINEERING COMMAND (Preparing Activity)

AIR FORCE CIVIL ENGINEER CENTER

Record of Changes (changes are indicated by \1\ ... /1/)

Change No.	Date	Location

This UFC supersedes UFC 4-030-01, dated December 2007 and UFC 3-400-01, dated July 2002 including Change 4, dated August 2008.

FOREWORD

The Unified Facilities Criteria (UFC) system is prescribed by MIL-STD 3007 and provides planning, design, construction, sustainment, restoration, and modernization criteria, and applies to the Military Departments, the Defense Agencies, and the DoD Field Activities in accordance with USD (AT&L) Memorandum dated 29 May 2002. UFC will be used for all DoD projects and work for other customers where appropriate. All construction outside of the United States is also governed by Status of Forces Agreements (SOFA), Host Nation Funded Construction Agreements (HNFA), and in some instances, Bilateral Infrastructure Agreements (BIA.) Therefore, the acquisition team must ensure compliance with the most stringent of the UFC, the SOFA, the HNFA, and the BIA, as applicable.

UFC are living documents and will be periodically reviewed, updated, and made available to users as part of the Services' responsibility for providing technical criteria for military construction. Headquarters, U.S. Army Corps of Engineers (HQUSACE), Naval Facilities Engineering Command (NAVFAC), and Air Force Civil Engineer Center (AFCEC) are responsible for administration of the UFC system. Defense agencies should contact the preparing service for document interpretation and improvements. Technical content of UFC is the responsibility of the cognizant DoD working group. Recommended changes with supporting rationale should be sent to the respective service proponent office by the following electronic form: Criteria Change Request. The form is also accessible from the Internet sites listed below.

UFCs are effective upon issuance and are distributed only in electronic media from the following source:
- Whole Building Design Guide web site http://dod.wbdg.org/.

Hard copies of UFC printed from electronic media should be checked against the current electronic version prior to use to ensure that they are current.

AUTHORIZED BY:

JAMES C. DALTON, P.E.
Chief, Engineering and Construction
U.S. Army Corps of Engineers

JOSEPH E. GOTT, P.E.
Chief Engineer
Naval Facilities Engineering Command

SCOTT HARTFORD, Colonel, USAF, P.E.
Acting Director
Facilities Engineering Center of Excellence
AF Civil Engineer Center

MICHAEL McANDREW
Director, Facilities Investment and Management
Office of the Deputy Under Secretary of Defense
(Installations and Environment)

UNIFIED FACILITIES CRITERIA (UFC)
NEW SUMMARY SHEET

Document: UFC 1-200-02, High Performance and Sustainable Building Requirements

Superseding: UFC 4-030-01, Sustainable Development and UFC 3-400-01, Energy Conservation

Description: This UFC was developed to bring uniformity across the DOD Components and:

- Drive transformation in the performance of the DOD facility inventory
- Require greater energy efficiency and water conservation measures that demonstrate a strong return on investment
- Balance building performance with occupant comfort, health, safety and productivity
- Cover planning, designing and acquiring new construction, existing facilities and infrastructure, leased space, and installations
- Guide compliance with higher level mandates, policies and standards
- Provide companion document to UFC 1-200-01, *General Building Requirements*

Reasons for Document: In an environment of diminishing resources, this document provides minimum unified requirements and coordinating guidance for planning, designing, constructing, renovating, and maintaining, high performance and sustainable facilities that will enhance DOD mission capability by reducing total ownership costs.

Impact: Improved mission capability through:

- Reduced total ownership costs of facilities
- Improved energy efficiency and water conservation
- Enhanced facility and installation performance and sustainability
- Promoting sustainable resource and environmental stewardship
- Enhance energy and water security

Unification Issues:

- The Navy and Air Force, in accordance with 10 CFR Part 433, require meeting the requirements of ASHRAE 90.1-2007. The Army requires meeting the requirements of ASHRAE 90.1-2010.

- The Army will utilize ASHRAE 189.1 Section 7.3.2 (On-Site Renewable Energy Systems) for future installation of on-site renewable energy systems, if on-site renewable energy systems are not life-cycle cost effective for new construction/major renovation design. In general, the Navy and Air Force will utilize an installation-level solution to renewable energy systems and will not require Section 7.3.2 if a building-level solution is not life cycle cost effective.

TABLE OF CONTENTS

TABLES

CHAPTER 1 INTRODUCTION

1-1 BACKGROUND.

The *Energy Policy Act of 2005* and Executive Orders (EO) 13423 and 13514 mandate Federal agencies to lead by example by promoting sustainable Federal infrastructure through environmentally-sound, economically-sound, and fiscally sound design, construction, and operating decisions. Consistent with UFC program requirements, this document integrates unique DOD requirements with Federal mandates and industry standards for high performance and sustainable buildings.

1-2 PURPOSE AND SCOPE.

This UFC provides minimum requirements and guidance to achieve high performance and sustainable facilities that comply with the *Energy Policy Act of 2005*, the *Energy Independence and Security Act of* 2007, EO 13423, EO 13514, and the *Guiding Principles for Federal Leadership in High Performance and Sustainable Buildings* (Guiding Principles). This UFC is organized around the Guiding Principles.

ASHRAE 189.1 is an industry standard created to consolidate and address sustainability and energy requirements. This UFC incorporates the sections of ASHRAE 189.1 mentioned herein when appropriate and determined to be life cycle cost effective. Where the provisions of ASHRAE 189.1 meet the intent of the Guiding Principles, the provisions of ASHRAE 189.1 are referenced as a means of compliance or provided as an alternative requirement. When other standards are referenced in ASHRAE 189.1, use the publication date of the standard referenced in Appendix A of this document.

1-3 ORGANIZATION.

Chapter 2 provides requirements for new construction and major renovations to existing buildings. Chapter 3 identifies the requirements for minor renovations to existing buildings. Chapter 4 details the federal requirements that must be met in order for an existing building to be considered a High Performance and Sustainable Building (HPSB). Chapter 5 provides HPSB reporting and third-party certification procedures. Chapter 6 highlights the more holistic vision of sustainability at the Installation level.

1-4 APPLICABILITY.

This UFC applies to all planning, design and construction, renovation, repair, maintenance and operation, and equipment installation in new and existing facilities and installations, regardless of funding source. Table 1-1 is applicable to all methods of project delivery and levels of construction as defined in UFC 1-200-01. This UFC also applies to the following construction activities to the greatest extent practical:

- Overseas facilities, considering mission objectives and Host Nation Agreements.

- Facilities supporting contingency operations.

- Non-permanent facilities.

1-5 GENERAL REQUIREMENTS.

UFC 1-200-01 provides applicability of model building codes and government-unique criteria for typical design disciplines and building systems, as well as for accessibility, antiterrorism, security, sustainability and safety. UFC 1-200-02 must be used with UFC 1-200-01 and the UFCs and government criteria referenced therein.

If any conflict occurs between this UFC and UFC 3-600-01, the requirements in 3-600-01 take precedence. In case of conflict between industry standards and military criteria, use the military requirements.

Comply with Table 1-1 Project/Work Type Compliance.

Table 1-1 Project/Work Type Compliance

PROJECT/WORK TYPE	UFC APPLICABILITY	COMPLY WITH CHAPTERS					
		1	2	3	4	5	6
New Construction	All	X	X			X	X
New Addition	Added SF and affected (existing) building systems[1]	X	X			X	X
Major Renovation[2]	>$2.5M[3] and >50% PRV[4]	X	X			X	X
Minor Renovation	<$2.5M[3] and/or <50% PRV[4], affected building systems[1]	X		X		X	X
O&M, Sustainment, Restoration and Modernization	Affected building systems[1]	X		X		X	X
Leased Buildings[5]	All	X			X	X	X

[1] "Affected building systems" – Projects/work must address the Federal requirements relevant to the existing building systems included in the scope of the project/work.
[2] Major Renovation includes changes to a building that provide significant opportunities to improve energy and water efficiency, which may include but are not limited to HVAC, lighting, building envelope and other building components that have a major impact on energy and water use. These may also include projects that provide an opportunity for compliance with other requirements in this UFC.
[3] In 2007 dollars, adjusted for inflation (utilize UFC 3-701-01, Table 4-2 "Military Construction Escalation Rates").
[4] Replacement cost is exclusive of the cost to bring the building into compliance.
[5] Chapter for Leased Buildings will be developed at a later date.

1-6   OTHER REQUIREMENTS.

Compliance, analysis and reporting requirements found in Chapter 5 apply to the entire DOD portfolio. See Appendix D for project requirements impacted by historic buildings, historic districts and those near historic facilities, view sheds and other designated cultural resources.

Work associated with meeting the requirements in this UFC must comply with USC work classification law.

1-7   LIFE CYCLE COST ANALYSIS (LCCA).

Life cycle cost effectiveness as defined in 10 CFR 433.2, applies to this entire document unless otherwise stated. All Life Cycle Cost Analyses (LCCA) performed must be prepared in accordance with 10 CFR Part 436, Subpart A and NIST Handbook 135 *"Life-Cycle Costing Manual for the Federal Energy Management Program"*. LCCA must be prepared using the Building Life Cycle Costing (BLCC) program, available from the National Institute of Standards and Technology (http://www.nist.gov/el/buildeconomic.cfm). A link to BLCC can also be found at the Department of Energy's building energy tools web site http://www1.eere.energy.gov/femp/information/download_blcc.html. When needed, use weather data obtained from UFC 3-400-02.

Any building-level LCCA must be calculated using a maximum of a 40-year building life and equipment lives based on accepted industry averages or as determined by the Authority Having Jurisdiction (AHJ). Individual components or systems life expectancies must be reflected by inclusion of appropriate replacement and salvage values at the appropriate year of the analysis.

LCCA's comparing individual component or system alternatives must use the estimated life of the mutually exclusive alternative having the longest life, not to exceed 40 years from the beginning of beneficial use or the lowest common multiple of the expected lives of the alternatives. Include the appropriate replacement and salvage values for each of the other alternatives. A number of tools comparing energy system alternatives and evaluating specific conservation measures may be of benefit in performing energy conservation calculations. Additional information can be found at: http://apps1.eere.energy.gov/buildings/tools_directory/.

All analysis must be performed based on the actual conditions expected over the life of the facility including anticipated occupancies, scheduled hours of operation and process loads. Realistic energy usage and efficiencies, maintenance cost and repairs and renovations must be included. All costs or savings associated with the utilization of recovered energy, solar heat, solar photovoltaic energy and other renewable or waste heat applications must be included. Any alternative funding such as rebates from utility companies should be appropriately credited in the LCCA.

1-8   REFERENCES.

Appendix A contains a list of references used in this document.

1-9   GLOSSARY.

Appendix C contains acronyms, abbreviations, and terms.

1-10   COMPLIANCE EXEMPTIONS

Refer to Chapter 5 for High Performance Sustainable Building compliance and third-party certification.

CHAPTER 2 NEW CONSTRUCTION AND MAJOR RENOVATION PROJECTS

2-1   OVERVIEW

The requirements of this chapter apply to all New Construction and Major Renovation projects. See Appendix D for project requirements impacted by historic buildings, historic districts and those near historic facilities, view sheds and other designated cultural resources.

Incorporate the requirements of this chapter in accordance with the following paragraphs of Chapter 1 in this UFC:

- Organization
- Applicability
- General Requirements
- Other Requirements
- Life Cycle Cost Analysis (LCCA)

2-2   EMPLOY INTEGRATED DESIGN PRINCIPLES.

Integrated design is the most important requirement in achieving a high performance building. A design team must have strong, consistent representation from all stake-holders throughout the project phases to avoid missing opportunities to improve building performance and to fully realize increase savings potential.

2-2.1   Integrated Design.

Use a collaborative, integrated planning and design team, composed of user, government support staff, and appropriate professionals, to identify requirements and to establish performance goals for siting, energy, water, materials, indoor environmental quality, and other comprehensive design goals. Ensure incorporation of these goals throughout the design and lifecycle of the building, including deconstruction.

Site attributes, including climate and local and regional context, impact the design of the building. Evaluate the site and building components to determine whether passive and natural design strategies and features are cost effectively incorporated before the active and mechanical systems are designed.

Use the following sequential approach:

- Analyze potential sites and associated climate including local and regional context.
- Select a sustainable site.
- Design building to site analysis focusing on orientation, configuration and massing.

- Select and integrate into the design, high performance and sustainable building envelope strategies and components (e.g. roofs, walls, fenestrations).

- Select, design and integrate into the overall design, high performance and sustainable systems (e.g. HVAC, plumbing, water heating systems, lighting systems, control systems, elevators, and fire protection systems).

Incorporate integrated design as described in ASHRAE 189.1 Informative Appendix H (Integrated Design). Follow the steps of design optimization, as applicable, in ASHRAE 189.1 Section H1.1.1 (Charrette Process).

2-2.2   Commissioning.

Employ commissioning practices appropriate to the size and complexity of the building and its system components in order to verify performance and ensure that design requirements are met. This must include an experienced commissioning provider. The use of contracted services or Government personnel as the commissioning provider will be determined at project level.

Meet the requirements of ASHRAE 189.1 Section 10.3.1.2 (Building Project Commissioning), with the following modifications:

- For buildings and systems that are less complex, commissioning can be tailored as determined by the DoD Component AHJ[6].

- "Schematic design" is the design charrette or similar conceptual design activity.

- Delete the requirement to meet ASHRAE 55 Section 6.2 (Documentation).

2-3   PROMOTE SUSTAINABLE LOCATION AND SITE DEVELOPMENT.

Sustainable site selection and site development are fundamental components of sustainable building practices. Opportunities offered by good site selection are available at the start of a project and in most cases incur no additional cost. Every opportunity should be taken to transform DOD installations into more livable, resource-efficient communities.

2-3.1   Site Selection.

Meet the requirements in UFC 2-100-01 and UFC 3-101-01.

During the site selection process, give preference to sites which possess the following characteristics when appropriate:

- Adaptive reuse or renovation of buildings or sites (including historic buildings), or use of previously disturbed/developed areas

---

[6] For Air Force projects, the Project Delivery Team must determine the level of commissioning activities required.

- Building orientation to maximize energy efficiency, passive solar and daylighting potential of the building

- Locations in central business districts (cantonment areas) and rural town centers (remote locations)

- Sites well served by public or other transit if available or projected to be available

- Site design elements that ensure safe and convenient pedestrian access

- Proximity to housing

- Avoiding development of sensitive land resources (including greenfields and USDA Prime Farmland as defined in USC Title 7 Section 4201)

- Consult with local Cultural Resource Manager to identify potential impacts to existing historic buildings, districts, or view sheds

- Parking management strategies to encourage co-use of shared parking

2-3.2   Mitigation of Heat Island Effect.

- For Site Hardscapes – Meet the requirements in ASHRAE 189.1 Section 5.3.2.1 (Site Hardscape) for sidewalks, courtyards and POV parking areas. For all other site hardscapes, implement to the maximum extent practical. The use of Section 5.3.2.1(e) is not allowed.

- For Walls – Meet the requirements in ASHRAE 189.1 Section 5.3.2.2 (Walls).

- For Roofs – Meet the requirements in UFC 3-110-03.

2-3.3   Reduction of Light Pollution.

Meet the requirements of UFC 3-530-01.

2-3.4   Stormwater Management.

Meet the requirements of UFC 3-210-10.

2-4   OPTIMIZE ENERGY PERFORMANCE.

The DoD objective of acquiring high performance and sustainable buildings is to reduce the total ownership costs of facilities. Building designs must be energy efficient, while balancing life-cycle costs, energy efficiency, energy security, and occupant benefits with budget and mission.

Meet the requirements of ASHRAE 90.1.[7]

Note: Title 10 CFR Part 433 requires all federal buildings for which design began on or after August 10, 2012 to (i) Meet ASHRAE 90.1–2007 and (ii) achieve energy

[7] For Navy and Air Force projects, use ASHRAE 90.1 (2007). For Army projects, use ASHRAE 90.1 (2010).

consumption levels that are at least 30-percent below the levels of the ASHRAE Baseline Building 2007 if life-cycle cost-effective. The Department of Energy determined that ASHRAE 90.1–2010 would achieve greater energy efficiency over ASHRAE 90.1–2007: site energy savings are estimated to be 18.5 percent (average).

2-4.1   Energy Efficiency.

Design the building to achieve at least 30% energy consumption reduction from ASHRAE 90.1 (2007) baseline or 12% energy consumption reduction from ASHRAE 90.1 (2010) baseline. [8]

Design Federal low-rise residential buildings to achieve at least 30% energy consumption reduction from International Energy Conservation Code (IECC) baseline.

If a 30% reduction is not life-cycle cost-effective, the design of the proposed building must be modified so as to achieve an energy consumption level at or better than the maximum level of energy efficiency that is life-cycle cost-effective.

2-4.2   On-Site Renewable Energy.

Provide on-site renewable energy systems in accordance with ASHRAE 189.1 Section 7.4.1.1 (On-Site Renewable Energy Systems) where life-cycle cost effective considering climate, infrastructure condition, mission compatibility, and effects on base wide electrical system (grid) power quality. Exception: Purchase of renewable energy certificates (RECs) must not be used as a substitute for the Section 7.4.1.1 new building requirement.

For Army projects, if not life-cycle cost effective, utilize ASHRAE 189.1 Section 7.3.2 (On-Site Renewable Energy Systems) for future installation of on-site renewable energy systems. [9]

When renewable systems are provided, comply with IEEE 1547 for renewable systems interface to the electrical system grid. *Note: The IEEE 1547 Series of Interconnection Standards are available from IEEE (http://grouper.ieee.org/groups/scc21/dr_shared/) as design guides for IEEE 1547 implementation.*

Designs must include the following requirements in order to limit the risk to energy security:

- a. Direct interconnection of system to installation-wide electrical system (grid):

  - For renewable systems that include a tie-in by a direct connection to the primary distribution system, provide a

---

[8] For Navy and Air Force projects, use ASHRAE 90.1 (2007) baseline. For Army projects, use ASHRAE 90.1 (2010) baseline.
[9] Navy and Air Force will utilize an installation-level solution to renewable energy systems and will not require ASHRAE 189.1 Section 7.3.2 if a building-level solution is not life cycle cost effective.

cumulative renewable load analysis of both the direct connection and building renewable power systems.

- Evaluate the proposed systems and verify there are no adverse effects on the installation-wide electrical system frequency control, voltage regulation and power quality. See Appendix B for issues that should be considered.

b. Direct interconnection of system to buildings that utilize engine generators that may operate in parallel with the renewable energy generation:

- Evaluate the proposed systems and verify there are no adverse effects on the generator's ability to maintain frequency control, voltage regulation and power quality. See Appendix B for issues that should be considered.

- If the stability of the combined systems cannot be confirmed, then design system to automatically disconnect renewable power generation when the backup generator is in operation.

c. Direct interconnection of system to buildings that either do not utilize engine generators or have backup power systems that would never operate in parallel with the renewable energy generation, such as a backup generator supplying power via an automatic transfer switch:

- Provide analysis validating that electrical system design address the electrical characteristics (real and reactive power output) of the renewable system.

Per EISA Section 523, meet at least 30% of the domestic hot water demand through the installation of solar water heating if life cycle cost effective.

Components may choose centralized renewable energy development in lieu of building by building application.

2-4.3   Energy Compliance Analysis.

Provide an Energy Compliance Analysis (ECA) which will demonstrate compliance with all energy related requirements in this UFC. Identify the specific energy conservation criteria that applies to the project, the software used to prepare the necessary calculations, a summary of all input to and output from the calculations, and the calculated baseline and as-designed building energy consumption of the proposed design. The ECA must include a completed "Energy Cost Budget (ECB) Compliance Report" as shown in Appendix G of the ASHRAE 90.1 "User's Manual."

Building-level energy consumption calculations must be performed using a professionally recognized and proven computer program or programs that integrate architectural features with air-conditioning, heating, lighting, and other energy producing or consuming systems. These programs will be capable of simulating the features, systems, and thermal loads used in the design. The energy savings and any parasitic energy loads associated with the utilization of recovered energy, solar heat, solar

9

photovoltaic energy and other renewable or waste heat applications must be included. The program will perform 8,760 hourly calculations. The Department of Energy maintains a list of building energy tools for design (such as eQuest by DoE).

A separate, concise LCCA narrative must be prepared by each of the following individuals: the Lead Project Architect, Lead Project Mechanical Engineer, and the Lead Project Electrical Engineer. Each of the three narratives must list the conservation features considered for that discipline, a list of conservation measures adopted in the design, and any supporting LCCA calculations. The results of this analysis are used for design decision-making in reducing total Life Cycle Cost, while meeting mission objectives.

The ECA must include all required LCCA studies and narratives. The ECA must be submitted for approval along with the Concept Design and updated as necessary as the design progresses.

2-4.4   Measurement and Verification.

A utility meter must be installed at each building, for each utility serving the building (e.g. district steam, district hot and chilled water, electricity, natural gas, fuel oil, etc.) in the normal units of the measure (i.e. kWh, kW, cf, gallons, etc.). Meters must be connected to a base wide energy and utility monitoring and control system using the installation's advanced metering protocols. The installation of meters is required per DODI 4170.11, Installation Energy Management.

Implement submetering when authorized in writing by the Installation. Utilize ASHRAE 189.1 Section 7.3.3 (Energy Consumption Management) as guide for subsystem implementation.

2-4.5   Benchmarking.

See Chapter 4, the paragraph entitled, "Benchmarking".

2-5   PROTECT AND CONSERVE WATER.

While the cost of water is generally lower than the cost of energy, conservation of water is no less important to the total cost of ownership of a facility. There are costs and energy expenditures associated with the production and delivery of water. As safe, clean water resources become scarce, the costs of water will increase. A new building is typically a 40-year decision and it must be made with water conservation in mind.

2-5.1   Indoor Water.

Meet the requirements of ASHRAE 189.1 Section 6.3.2.1 (Plumbing Fixtures and Fittings). Water closet replacements in major renovations may have a flush value of up to 1.6 GPF (6.1 LPF) to accommodate existing plumbing capacity.

Specify EPA WaterSense-labeled products, where appropriate and available.

Meet the requirements of ASHRAE 189.1 Section 6.3.2.2 (Appliances).

2-5.2 Outdoor Water.

Use water efficient landscape and irrigation strategies, such as water reuse, xeriscaping and the use of harvested rainwater, to reduce outdoor potable water consumption by a minimum of 50%. This requirement must be demonstrated by comparison to a baseline building using the EPA WaterSense landscape water budget tool version 1.01 or later, or a Component approved tool.

Meet the requirements of UFC 3-201-02 and the following requirements of ASHRAE 189.1 when life cycle cost effective: Section 6.3.1 (Mandatory Provisions: Site Water Use Reduction) and either Section 6.4.1 (Prescriptive Option: Site Water Use Reduction) or Section 6.5.1 (Performance Option: Site Water Use Reduction).

Show preference for irrigation contractors who are certified through a WaterSense labeled program.

2-5.3 Water Used for Heating and Cooling.

Meet the requirements of the following sections of ASHRAE 189.1 when life cycle cost effective.

- Meet the requirements of Sections 6.3.2.3(a) and (b) (HVAC Systems and Equipment).
- Section 6.3.2.3(c) (HVAC Systems and Equipment). Condensate collection is not required if there is no practical use for the collected condensate.
- Section 6.4.2.1 (Cooling Towers).

Per EPAct 2005 Section 109, when potable water is used to improve a building's energy efficiency, implement lifecycle cost effective water conservation measures.

2-5.4 Measurement of Water Use.

Install water meters to monitor use in buildings consuming indoor and outdoor water as required by DODI 4170.11.

The installation of separate water meters for locations with outdoor water use is encouraged.

2-6 ENHANCE INDOOR ENVIRONMENTAL QUALITY.

Mission and User requirements, as well as occupant health and safety all influence building performance and must be achieved when selecting energy efficient options during design, construction and operations.

2-6.1 Ventilation and Thermal Comfort.

Meet the requirements of ASHRAE 189.1 Section 8.3.1 (Indoor Air Quality) and Section 8.3.2 (Thermal Environmental Conditions for Human Occupancy). Documentation as described in ASHRAE 55 Section 6.2 is not required. Compliance with standards by providing passive (non-mechanical) thermal comfort methods are allowed and encouraged as described in the paragraph entitled, "Integrated Design" in this UFC.

2-6.2 Moisture Control.

Establish and implement a moisture control strategy for controlling moisture flows and condensation to prevent building damage, minimize mold contamination, and reduce health risks related to moisture. Meet the requirements of ASHRAE 189.1 Section 10.3.1.5 (Moisture Control) and UFC 3-101-01 Chapter 3, Building Envelope Requirements.

2-6.3 Daylighting.

Meet the requirements of ASHRAE 189.1 Section 8.4.1 (Daylighting by Sidelighting) or Section 8.5.1 (Daylighting Simulation). Provide automated lighting controls in accordance with UFC 3-530-01.

2-6.4 Low-Emitting Materials.

Specify materials and products with low pollutant emissions, including composite wood products, adhesives, sealants, interior paints and finishes, carpet systems, and furnishings. Meeting the requirements of ASHRAE 189.1 Sections 8.4.2 (Prescriptive Option: Materials) or Section 8.5.2 (Performance Option: Materials) demonstrates compliance.

2-6.5 Protect Indoor Air Quality during Construction.

Follow the recommended approach of SMACNA 008. After construction and prior to occupancy, conduct a minimum 72-hour flush-out with maximum outdoor air consistent with achieving relative humidity no greater than 60%. After occupancy, continue flush-out as necessary to minimize exposure to contaminants from new building materials. Meet the requirements of ASHRAE 189.1 Section 10.3.1.4 (Indoor Air Quality (IAQ) Construction Management).

2-7 REDUCE ENVIRONMENTAL IMPACT OF MATERIALS.

2-7.1 Environmentally Preferable Products.

Use products that have a lesser or reduced effect on human health and the environment. Consider product life-cycle when compared with competing products or services serving the same purpose. A number of standard and eco-labels are available in the marketplace to assist specifiers in making environmentally preferable decisions. Electronic devices, products and appliances purchased should be EPEAT certified. For a list of products and recommendations designated for Federal procurement, consult the Green Products Compilation at http://www.gsa.gov/greenproductscompilation and the *Federal Green Construction Guide for Specifiers* at http://www.wbdg.org/design/greenspec.php.

2-7.2   Recycled Content.

Meet the requirements of ASHRAE 189.1 Section 9.4.1.1 (Recycled Content) when products are available.

Per RCRA Section 6002, for EPA-designated products, specify products meeting or exceeding EPA's recycled content recommendations. EPA's recycled content product designations and recycled content recommendations are available on EPA's Comprehensive Procurement Guideline web site at http://www.epa.gov/epawaste/conserve/tools/cpg/index.htm. Requirements will be met when products are available.

2-7.3   Biologically-Based Products.

Per Section 9002 of the Farm Security and Rural Investment Act, specify products composed of the highest percentage of biobased content consistent with the USDA BioPreferred Program, if products meet performance requirements and are available at a reasonable cost. USDA's biobased product designations and biobased content recommendations are available on USDA's BioPreferred web site at http://www.biopreferred.gov/.

2-7.4   Waste and Materials Management.

Meet the requirements of ASHRAE 189.1 Section 9.3.1.1 (Construction Waste Management-Diversion) and Section 9.3.4.1 (Storage and Collection of Recyclables – Recyclables), where markets or onsite recycling opportunities exist or are anticipated.

2-7.5   Ozone Depleting Substances.

Meet the requirements of ASHRAE 189.1 Section 9.3.3 (Refrigerants). Exceptions: Refer to UFC 3-600-01 for fire system requirements.

This Page Intentionally Left Blank

CHAPTER 3 MINOR RENOVATION PROJECTS

3-1   OVERVIEW.

The requirements of this chapter apply to all projects with the exception of New Construction and Major Renovations. The goal is to improve the performance of the existing building inventory with every investment. Examples of projects that must comply with this chapter include repair, maintenance, and equipment installations. It is not expected that the requirements will be fully met by all projects; nor that all projects will address all the requirements. Project teams will set systems performance goals relative to 1) the project scope of work, 2) the intent of the Federal requirements, and 3) with the ultimate performance thresholds in mind. Projects must meet the requirements in this chapter relative to the scope for each project.

See Appendix D for project requirements impacted by historic buildings, historic districts and those near historic facilities, view sheds and other designated cultural resources.

Incorporate the requirements of this chapter in accordance with the following paragraphs of Chapter 1 of this UFC.

- Organization
- Applicability
- General Requirements
- Other Requirements
- Life Cycle Cost Analysis (LCCA)

3-2   EMPLOY INTEGRATED ASSESSMENT, OPERATION, AND MANAGEMENT PRINCIPLES.

3-2.1   Integrated Assessment, Operation, and Management.

Use a collaborative, integrated planning and design team composed of user, government support staff and appropriate professionals to identify requirements and to establish performance goals for energy, water, materials, indoor environmental quality, and other comprehensive design goals. Ensure incorporation of these goals throughout the design and lifecycle of the building, including deconstruction.

Evaluate the site and building components to determine whether passive and natural design strategies and features are cost effectively incorporated before the active and mechanical systems are designed.

- Select and integrate into the design, high performance and sustainable building envelope strategies and components (e.g. roofs, walls, fenestrations).
- Select, design and integrate into the overall design, high performance and sustainable systems (e.g. HVAC, plumbing, water heating systems, lighting systems, control systems, elevators, fire protection systems).

3-2.2   Commissioning, Re-Commissioning, and Retro-Commissioning.

Employ commissioning, tailored to the size and complexity of the existing building and its system components, in order to optimize and verify performance of fundamental building systems. Re-Commissioning or Retro-Commissioning must be performed by an experienced in-house or external provider. When building commissioning has been performed, the commissioning report and summary of actions taken must be documented.

3-3   PROMOTE SUSTAINABLE LOCATION AND SITE DEVELOPMENT.

3-3.1   Maximize Efficient Use of Existing Workplaces.

Before altering or adding to the building, evaluate options for and employ alternative space management and workplace arrangements.

3-3.2   Reduce Transportation Related Greenhouse Gas Emissions.

Before alteration or proposing an addition to a facility, apply the master planning principles in UFC 2-100-01 to address transportation considerations. This includes consideration for connected sidewalks and trails, bicycle paths and storage areas as well as other solutions that promote walkable neighborhoods.

3-3.3   Integrate with Local Planning.

Per Section 2(f) of EO 13514, engage with regional and local integrated plans for transportation and energy choices. For large facilities, with more than 100 employees, consult with local planning and transportation officials on an annual or biennial basis, as appropriate, to discuss opportunities to coordinate planning efforts.

3-3.4   Mitigate the Heat Island Effect.

When planning additions to or renovations of the building and the surrounding site, consider strategies such as trees and vegetation, vegetative roofs, highly reflective roofs, and cool hardscapes, to mitigate the heat island effect. Implementing measures in UFC 3-110-03 and ASHRAE 189.1 Section 5.3.2 (Mitigation of Heat Island Effect) are encouraged as appropriate. The use of Section 5.3.2.1(e) is not allowed.

3-3.5   Reduce Light Pollution.

Meet the requirements of UFC 3-530-01.

3-3.6   Stormwater.

Meet the requirements of UFC 3-210-10.

3-4   OPTIMIZE ENERGY PERFORMANCE.

3-4.1   Energy Efficiency.

The total building, minimum energy performance targets for all Federal facilities are listed below.

- Option 1: Receive an ENERGY STAR® performance score of 75 or higher.

- Option 2: Reduce measured building energy use by at least 20% compared to building energy use in 2003 or a year thereafter with metered energy use data.

- Option 3: Reduce modeled energy use (from all sources including renewable energy) by 20% compared to the ASHRAE 90.1 baseline building design.

Submit supporting LCCA documentation to include a concise narrative that lists the energy conservation features considered and adopted in the design and any supporting LCCA calculations.

Per EO 13514 and EISA Section 525, acquire products that are ENERGY STAR®-qualified or meet FEMP-designated efficiency requirements in all covered product categories. Select products based on life cycle cost, not initial cost. Follow EISA Section 524 requirements when purchasing devices that use or contain a standby power function.

3-4.2   On-Site Renewable Energy.

Implement renewable energy generation projects on agency property for agency use, where lifecycle cost effective and when there are no adverse effects on base wide electrical system (grid) power quality.

When renewable systems are provided, comply with IEEE 1547 for renewable systems interface to the electrical system grid. *Note: The IEEE 1547 Series of Interconnection Standards are available from IEEE (http://grouper.ieee.org/groups/scc21/dr_shared/) as design guides for IEEE 1547 implementation.*

Designs must include the following requirements in order to limit the risk to energy security:

    a. Direct interconnection of system to base-wide electrical system:

- For renewable systems that include a tie-in by a direct connection to the primary distribution system, provide a cumulative renewable load analysis of both the direct connection and building renewable power systems.

- Evaluate the proposed systems and verify there are no adverse effects on the base-wide electrical system frequency control, voltage regulation and power quality. See Appendix B for issues that should be considered.

b. Direct interconnection of system to buildings that utilize engine generators that may operate in parallel with the renewable energy generation:

- Evaluate the proposed systems and verify there are no adverse effects on the generator's ability to maintain frequency control, voltage regulation and power quality. See Appendix B for issues that should be considered.

- If the stability of the combined systems cannot be confirmed, then design system to automatically disconnect renewable power generation when the backup generator is in operation.

c. Direct interconnection of system to buildings that either do not utilize engine generators or have backup power systems that would never operate in parallel with the renewable energy generation, such as a backup generator supplying power via an automatic transfer switch:

- Provide analysis validating that electrical system design address the electrical characteristics (real and reactive power output) of the renewable system.

Components may choose centralized renewable energy development in lieu of building by building application.

3-4.3 Measurement and Verification.

A utility meter must be installed at each building, for each utility serving the building (e.g. district steam, district hot and chilled water, electricity, natural gas, fuel oil, etc.) in the normal units of the measure (i.e. kWh, kW, cf, gallons, etc.). All meters must be connected to a base wide energy and utility monitoring and control system using the installation's advanced metering protocols. The installation of meters is required per DODI 4170.11.

Implement submetering when authorized in writing by the installation.

3-4.4 Benchmarking.

See Chapter 4, the paragraph entitled, "Benchmarking".

3-5 PROTECT AND CONSERVE WATER.

3-5.1 Indoor Water.

Use one of the two following options to reduce indoor potable water use reduction:

- Option 1: Reduce building potable water use to a level equal to or lower than a baseline calculated assuming 100% of the plumbing fixtures and fittings meet the Uniform Plumbing Code (UPC) 2006 or the International

Plumbing Code (IPC) 2006 fixture and fitting performance requirements. The baseline for plumbing fixtures and fittings installed before 1994 is allowed to be calculated at 140% of the UPC or IPC performance requirements, or

- Option 2: Reduce building measured potable water use by 20% compared to building water use in 2003 or a year thereafter with metered water data. If a building has a single water meter, reduce indoor and outdoor potable water use combined by at least 20% compared to building water use in 2003 or a year thereafter with metered water data.

3-5.2   Outdoor Water.

Use of potable water for landscape irrigation is not desired, particularly in areas of high cost or low availability. Use one of the three following three options to reduce outdoor potable water use for landscaping:

- Option 1: Reduce potable irrigation water use by 50% compared to the conventional methods, or

- Option 2: Reduce building related potable irrigation water use by 50% compared to irrigation water use in 2003 or a year thereafter with metered water data. If a building has a single water meter, reduce indoor and outdoor potable water use combined by at least 20% compared to building water use in 2003 or a year thereafter with metered water data, or

- Option 3: Use no potable irrigation water.

3-5.3   Water Used for Heating and Cooling.

Per EPAct 2005 Section 109, when potable water is used to improve a building's energy efficiency, deploy lifecycle cost effective water conservation measures.

3-5.4   Measurement of Water Use.

Install water meters to monitor use in buildings consuming indoor and outdoor water as required by DODI 4170.11.

3-5.5   Water-Efficient Products and Services.

Where appropriate and available, use EPA WaterSense-labeled products. For product categories not covered by WaterSense, use the most water-efficient product practical. Show preference for irrigation contractors who are certified through a WaterSense-labeled program.

3-6   ENHANCE INDOOR ENVIRONMENTAL QUALITY.

3-6.1   Ventilation and Thermal Comfort.

At a minimum, meet ASHRAE 55 and ASHRAE 62.1. Documentation as described in ASHRAE 55 Section 6.2 is not required.

3-6.2   Moisture Control.

Establish and implement a moisture control strategy for controlling moisture flows and condensation to prevent building damage, minimize mold contamination, and reduce health risks related to moisture. Meet the requirements of ASHRAE 189.1 Section 10.3.1.5 (Moisture Control) and UFC 3-101-01 Chapter 3, Building Envelope Requirements.

3-6.3   Daylighting and Lighting Controls.

Use these two strategies in any combination to meet daylighting and lighting controls performance:

- Strategy 1: Achieve a minimum daylight illumination level of 25 foot-candles (269 lux) in 50% of all regularly occupied spaces, or

- Strategy 2: Provide occupant controlled lighting, allowing adjustments to suit individual task needs, for 50% of regularly occupied offices and work stations.

Provide automated lighting controls in accordance with UFC 3-530-01.

3-6.4   Low-Emitting Materials.

Use low emitting materials for building modifications, maintenance, and cleaning. In particular, specify the following materials and products to have low pollutant emissions: composite wood products, adhesives, sealants, interior paints and finishes, solvents, carpet systems, janitorial supplies, and furnishings. Meeting the requirements of ASHRAE 189.1 Sections 8.4.2 (Prescriptive Option: Materials) or Section 8.5.2 (Performance Option: Materials) demonstrates compliance. Requirements will be met unless installation location, mission and facility type limit compliance, or practical application such as industrial spaces or other unique facility conditions limits compliance.

3-7   REDUCE ENVIRONMENTAL IMPACT OF MATERIALS.

3-7.1   Environmentally Preferable Products.

Use products that have a lesser or reduced effect on human health and the environment over their lifecycle when compared with competing products or services that serve the same purpose. A number of standards and eco-labels are available in the marketplace to assist specifiers in making environmentally preferable decisions. For a list of products and recommendations designated for Federal procurement, consult the Green Products Compilation at http://www.gsa.gov/greenproductscompilation and the *Federal Green Construction Guide for Specifiers* at http://www.wbdg.org/design/greenspec.php.

3-7.2   Recycled Content.

Per RCRA Section 6002, meet or exceed the minimum recycled content recommendations in solicitations for purchase of EPA-designated products for building modifications, operations and maintenance, and cleaning, if products meet performance requirements and are available at a reasonable cost. A preference for purchasing EPA-designated products must be included in all applicable solicitations. EPA's recycled content product designations and recycled content recommendations are available on EPA's Comprehensive Procurement Guideline web site at http://www.epa.gov/epawaste/conserve/tools/cpg/index.htm.

In building modification projects, where products meet performance requirements and are available at a reasonable cost, use recycled content materials for which the post-consumer recycled content plus half of the pre-consumer content equals at least 10% of the total value of the materials (based on cost or weight).

3-7.3   Biologically-Based Products.

Per Section 9002 of the Farm Security and Rural Investment Act, specify products composed of the highest percentage of biobased content consistent with the USDA BioPreferred Program, if products meet performance requirements and are available at a reasonable cost. USDA's biobased product designations and biobased content recommendations are available on USDA's BioPreferred web site at http://www.biopreferred.gov/.

Use certified sustainable wood products, where products meet performance requirements and are available at a reasonable cost. Requirements will be met unless installation location, mission and facility type limit compliance or practical application.

3-7.4   Waste and Materials Management.

Provide space for reuse, salvage and recycling services for building occupants, where markets or on-site recycling exist.

Provide salvage, reuse and recycling services for waste generated from construction, minor renovations, repairs, maintenance and equipment installations, where markets exist.

3-7.5   Ozone Depleting Substances.

Meet the requirements of ASHRAE 189.1 Section 9.3.3 (Refrigerants) if equipment or systems using ozone depleting substances are included in the project. Exception: Refer to UFC 3-600-01 for fire system requirements.

This Page Intentionally Left Blank

CHAPTER 4 HIGH PERFORMANCE AND SUSTAINABLE BUILDING (HPSB) ASSESSMENT REQUIREMENTS FOR EXISTING BUILDINGS

4-1 OVERVIEW.

This chapter details the Federal assessment requirements that must be met in order for an existing building to be considered a High Performance and Sustainable Building (HPSB). HPSB reporting procedures are covered in the paragraph entitled, "Compliance Reporting".

4-2 EMPLOY INTEGRATED ASSESSMENT, OPERATION, AND MANAGEMENT PRINCIPLES.

4-2.1 Integrated Assessment, Operation, and Management.

Establish an integrated team to develop and implement practices regarding building-level sustainable operations and maintenance.

- Incorporate sustainable operations and maintenance practices within the appropriate Environmental Management System (EMS).

- Assess existing condition and operational procedures of the building and major building systems (See the paragraph entitled, "Compliance Strategies for Aligning Investments with Sustainability") and identify areas for improvement, including evaluating options for alternatives for space management and workplace arrangements.

- Establish operational performance goals for energy, water, material use and recycling, and indoor environmental quality, and ensure incorporation of these goals throughout the remaining lifecycle of the building.

- Incorporate a building management plan to ensure that operating decisions and tenant education are carried out with regard to integrated, sustainable building operations and maintenance.

- Augment building operations and maintenance as needed using occupant feedback on workspace satisfaction.

4-2.2 Re-Commissioning and Retro-Commissioning.

Employ commissioning, tailored to the size and complexity of the existing building and its system components, in order to optimize and verify performance of fundamental building systems. Re-Commissioning or Retro-Commissioning must be performed by an experienced in-house or external provider. When building commissioning has been performed, the commissioning report and summary of actions taken must be documented.

4-3   PROMOTE SUSTAINABLE LOCATION AND SITE DEVELOPMENT.

4-3.1   Reduce Transportation-Related Greenhouse Gas Emissions.

Disseminate information to building occupants about available transportation alternatives, amenities within walking distance, and alternative workplace arrangements.

4-3.2   Integrate with Local Planning.

Per Section 2(f) of EO 13514, engage with regional and local integrated plans for transportation and energy choices. For large facilities, with more than 100 employees, consult with local planning and transportation officials on an annual or biennial basis, as appropriate, to discuss opportunities to coordinate planning efforts.

4-4   OPTIMIZE ENERGY PERFORMANCE.

4-4.1   Energy Efficiency.

Use one of the following three options to optimize and measure energy efficiency performance:

- Option 1: Receive an ENERGY STAR® performance score of 75 or higher.

- Option 2: Reduce measured building energy use by at least 20% compared to building energy use in 2003 or a year thereafter with metered energy use data.

- Option 3: Reduce modeled energy use (from all sources including renewable energy) by 20% compared to the ASHRAE 90.1 baseline building design.

Per EO 13514 and the EISA Section 525, acquire products that are ENERGY STAR®-qualified or meet FEMP-designated efficiency requirements in all covered product categories. Select products based on life cycle cost, not initial cost. Follow EISA Section 524 requirements when purchasing devices that use or contain a standby power function.

4-4.2   On-Site Renewable Energy.

Implement renewable energy generation projects on agency property for agency use, where lifecycle cost effective and when there are no adverse effects on base wide electrical system (grid) power quality.

When renewable systems are provided, comply with IEEE 1547 for renewable systems interface to the electrical system grid. *Note: The IEEE 1547 Series of Interconnection Standards are available from IEEE (http://grouper.ieee.org/groups/scc21/dr_shared/) as design guides for IEEE 1547 implementation.*

Designs must include the following requirements in order to limit the risk to energy security:

    a.   Direct interconnection of system to base wide electrical system:

- For renewable systems that include a tie-in by a direct connection to the primary distribution system, provide a cumulative renewable load analysis of both the direct connection and building renewable power systems.

- Evaluate the proposed systems and verify there are no adverse effects on the base-wide electrical system frequency control, voltage regulation and power quality. See Appendix B for issues that should be considered.

    b.   Direct interconnection of system to buildings that utilize engine generators that may operate in parallel with the renewable energy generation:

- Evaluate the proposed systems and verify there are no adverse effects on the generator's ability to maintain frequency control, voltage regulation and power quality. See Appendix B for issues that should be considered.

- If the stability of the combined systems cannot be confirmed, then design system to automatically disconnect renewable power generation when the backup generator is in operation.

    c.   Direct interconnection of system to buildings that either do not utilize engine generators or have backup power systems that would never operate in parallel with the renewable energy generation, such as a backup generator supplying power via an automatic transfer switch:

- Provide analysis validating that electrical system design address the electrical characteristics (real and reactive power output) of the renewable system.

Components may choose centralized renewable energy development in lieu of building by building application.

4-4.3   Measurement and Verification.

Install a utility meter for each utility serving the building (e.g. district steam, district hot and chilled water, electricity, natural gas, fuel oil, etc.) in the normal units of the measure (i.e. kWh, kW, cf, gallons, etc.). Connect all meters to a base wide energy and utility monitoring and control system using the installation's advanced metering protocols. The installation of meters is required per DODI 4170.11.

4-4.4   Benchmarking.

Compare annual performance data with previous years' performance data, by entering annual performance data in a measurement and tracking tool such as EPA's ENERGY STAR®, Portfolio Manager. The Labs21® Energy Benchmarking Tool may be used to benchmark laboratory buildings. Results of benchmarking should be used to inform capital investment decisions for future projects.

4-5   PROTECT AND CONSERVE WATER.

4-5.1   Indoor Water.

Use one of the two following options to reduce indoor potable water use reduction:

- Option 1: Reduce building potable water use to a level equal to or lower than a baseline calculated assuming 100% of the plumbing fixtures and fittings meet the Uniform Plumbing Code (UPC) 2006 or the International Plumbing Code (IPC) 2006 fixture and fitting performance requirements. The baseline for plumbing fixtures and fittings installed before 1994 is allowed to be calculated at 140% of the UPC or IPC performance requirements, or

- Option 2: Reduce building measured potable water use by 20% compared to building water use in 2003 or a year thereafter with metered water data. If a building has a single water meter, reduce indoor and outdoor potable water use combined by at least 20% compared to building water use in 2003 or a year thereafter with metered water data.

4-5.2   Outdoor Water.

Use of potable water for landscape irrigation is not desired, particularly in areas of high cost or low availability. Use one of the three following three options to reduce outdoor potable water use for landscaping:

- Option 1: Reduce potable irrigation water use by 50% compared to the conventional methods, or

- Option 2: Reduce building related potable irrigation water use by 50% compared to irrigation water use in 2003 or a year thereafter with metered water data. If a building has a single water meter, reduce indoor and outdoor potable water use combined by at least 20% compared to building water use in 2003 or a year thereafter with metered water data, or

- Option 3: Use no potable water for irrigation purposes.

4-5.3   Water Used for Heating and Cooling.

Per EPAct 2005 Section 109, when potable water is used to improve a building's energy efficiency, deploy lifecycle cost effective water conservation measures.

4-5.4   Measurement of Water Use.

Install water meters to monitor use in buildings consuming indoor and outdoor water as required by DODI 4170.11.

4-6 ENHANCE INDOOR ENVIRONMENTAL QUALITY.

4-6.1 Ventilation and Thermal Comfort.

At a minimum, meet ASHRAE 55 and ASHRAE 62.1. Documentation as described in ASHRAE 55 Section 6.2 is not required.

4-6.2 Moisture Control.

Establish and implement a moisture control strategy for controlling moisture flows and condensation to prevent building damage, minimize mold contamination, and reduce health risks related to moisture. Meet the requirements of ASHRAE 189.1 Section 10.3.1.5 (Moisture Control) and UFC 3-101-01 Chapter 3, Building Envelope Requirements.

4-6.3 Daylighting and Lighting Controls.

Use these two strategies in any combination to meet daylighting and lighting controls performance:

- Strategy 1: Achieve a minimum daylight illumination level of 25 foot-candles (269 lux) in 50 percent of all regularly occupied spaces, or

- Strategy 2: Provide occupant controlled lighting, allowing adjustments to suit individual task needs, for 50 percent of regularly occupied offices and work stations.

Provide automated lighting controls in accordance with UFC 3-530-01.

4-6.4 Low-Emitting Materials.

Use low emitting materials for building modifications, maintenance, and cleaning. In particular, specify the following materials and products to have low pollutant emissions: composite wood products, adhesives, sealants, interior paints and finishes, solvents, carpet systems, janitorial supplies, and furnishings.

4-6.5 Integrated Pest Management.

Use integrated pest management techniques as appropriate to minimize pesticide usage. Use EPA-registered pesticides only when needed.

4-6.6 Environmental Tobacco Smoke Control.

Prohibit smoking within the building and within 25 feet (7.62 meters) of all building entrances, operable windows, and building ventilation intakes.

4-7   REDUCE ENVIRONMENTAL IMPACT OF MATERIALS.

4-7.1   Environmentally Preferable Products.

Use products that have a lesser or reduced effect on human health and the environment over their lifecycle when compared with competing products or services that serve the same purpose. A number of standards and eco-labels are available in the marketplace to assist specifiers in making environmentally preferable decisions. For a list of products designated for Federal procurement, consult the Green Products Compilation at www.gsa.gov/greenproductscompilation.

4-7.2   Recycled Content.

Per Section 6002 of the Resource Conservation and Recovery Act (RCRA), meet or exceed the minimum recycled content recommendations when using EPA-designated products for tenant agency fit-out, maintenance, and cleaning, if products meet performance requirements and are available at a reasonable cost. EPA's recycled content product designations and recycled content recommendations are available on EPA's Comprehensive Procurement Guideline web site at www.epa.gov/cpg.

Encourage reuse of building materials, components, and furnishings that are in refurbishable condition and meet the quality standards of the government.

4-7.3   Biologically-Based Products.

Per Section 9002 of the Farm Security and Rural Investment Act, use products composed of the highest percentage of biobased content consistent with the USDA BioPreferred Program, if products meet performance requirements and are available at a reasonable cost. A preference for purchasing products with the highest biobased content per USDA recommendations for designated product categories must be included in all applicable solicitations for building modifications, operations and maintenance, and cleaning. USDA's biobased product designations and biobased content recommendations are available on USDA's BioPreferred website at http://www.biopreferred.gov/.

Use certified sustainable wood products, where products meet performance requirements and are available at a reasonable cost.

4-7.4   Waste and Materials Management.

Provide recycling services for paper, corrugated cardboard, glass, plastics, and metals, at a minimum, where markets or on-site recycling exist.

Provide salvage, reuse and recycling services for waste generated from building operations, maintenance, repair and minor renovations, where markets or on-site recycling exist. Divert a minimum of 50% of construction waste.

4-7.5   Ozone Depleting Substances.

Eliminate the use of ozone depleting substances (ODS) by 2020 where substitutes have been found acceptable for use by EPA's SNAP program. Refer to the list of substitutes for ODS that have been found acceptable for relevant end uses at www.epa.gov/ozone/snap/. Exception: Refer to UFC 3-600-01 for fire protection requirements.

This Page Intentionally Left Blank

CHAPTER 5 FEDERAL HPSB COMPLIANCE, REPORTING, AND SUSTAINABILITY POLICIES

5-1   BACKGROUND.

The first sustainable Federal building performance metric addressed energy efficiency and targeted new construction only [Energy Policy Act of 2005, (EPAct 05)]. Since EPAct 05, the scope of the building performance requirements has expanded to other areas of sustainability; and applies, in varying degrees, to the entire Federal building portfolio. The Federal requirements collectively are referred to as the Guiding Principles for Federal High Performance and Sustainable Buildings (HPSB). EO 13423 requires 1) new construction and major renovation, and 2) by 2015, 15% of the existing building inventory to comply with the requirements mentioned herein. EO 13514 clarifies the Guiding Principles apply to buildings larger than 5,000 SF (465 SM); and requires Federal agencies to show annual progress beyond 15%, with the ultimate goal of 100% of the building inventory in compliance.

5-2   COMPLIANCE WITH FEDERAL REQUIREMENTS.

This UFC has been designed to consolidate federal requirements for High Performance and Sustainable Buildings. The requirements herein must be addressed in order for a new construction/major renovation project or minor construction to an existing building, to be considered in compliance.

Meeting the intent of the Guiding Principles in some instances may be addressed at the installation level versus the building level (e.g. no smoking policy, storm water management). For guidance on installation-wide Guiding Principles refer to Chapter 6.

5-2.1   Compliance Checklists.

DoD Component-level HPSB checklists provide a method for documenting whether or not a facility meets the requirements to be a High Performance and Sustainable Building.

- Air Force – *Air Force MILCON Sustainability Requirements Scoresheet*, Attachment 3 of "AF Sustainable Design Development Implementing Guidance Memo" (http://www.wbdg.org/ccb/browse_cat.php?c=265)

- Army – *ENERGY & SUSTAINABILITY RECORD CARD*

- Navy – *NAVFAC Sustainability & ENERGY DATA RECORD CARD* (http://www.wbdg.org/pdfs/navfac_sustainable_energy_data_record_card.pdf)

5-2.2   New Construction and Major Renovations.

A building meeting the new construction and major renovations requirements also meets the Guiding Principles requirements for existing buildings and can contribute to the requirement mentioned in the paragraph entitled, "Background" in this UFC. For reporting requirements, refer to the paragraph entitled, "Compliance Reporting".

5-2.3   Minor Renovations to Existing Buildings.

For reporting requirements, refer to the paragraph entitled, "Compliance Reporting".

5-2.4   Existing Buildings, HPSB Assessment.

In a DoD installation setting, as much as 75% of the HPSB requirements for existing buildings are addressed by successful, fully mature, installation-wide policies, programs, contracts, and specifications. Generally, the "installation-wide" Guiding Principles response (to a specific requirement) will be the same for all buildings on an installation.

Once a building has been assessed to be in compliance with the Guiding Principles, it can be assumed to be in compliance for no more than five years, after which a re-assessment must be conducted in order to determine the compliance status.

For reporting requirements, refer to the paragraph entitled, "Compliance Reporting".

5-2.5   Identification and Documentation of "Not Applicable" Requirements.

The focus of compliance with the Guiding Principles is to meet the requirements applicable to a given building/location/mission. Compliance is demonstrated by meeting all of the applicable requirements, and documenting when certain requirements are not applicable (such as daylighting in a theater).

5-2.6   Compliance Strategies for Aligning Investments with Sustainability.

DoD Components are encouraged to align similar HPSB and energy and water activities to identify projects that: 1) address both sets of requirements, 2) are prioritized based on return on investment, and 3) offer whole building, maintainable, energy and water efficiency solutions. Selection of a single data management tool will afford the highest level of integration of the two programs.

As mentioned in the paragraph entitled, "Existing Buildings, HPSB Assessment", the majority of the existing building assessment requirements are generally addressed by installation-wide policies, programs, contracts, and specifications found in Chapter 6. The remaining 25% of the Guiding Principles are building specific and linked to energy and water consumption. This linkage affords DoD the opportunity to leverage ongoing energy efficiency and water conservation activities to advance compliance with the HPSB requirement. Specifically, many of the EISA Section 432 energy and water management requirements align with one or more Guiding Principles. This list includes but is not limited to:

- Commissioning, Recommissioning, and Retrocommissioning
- Energy and Water Evaluations
- Identifying and Implementing Energy and Water Efficiency Measures
- Web-Based Tracking

5-3   COMPLIANCE REPORTING.

Reporting compliance for the DoD Components to OSD with the EO 13423 and EO 13514 HPSB requirements is accomplished through the Real Property Inventory Submission as provided in guidance promulgated by the Deputy Under Secretary of Defense (Installations & Environment), Business Enterprise Integration Directorate (BEI). Policy regarding this reporting requirement can be found in DODI 4165.14.

BEI maintains the real property inventory data standards needed for the Enterprise Energy Information Management capability in the Real Property Information Model. The specific data element identified with this reporting requirement is "RPA Sustainability Code". The Air Force[10], Army[11], and Navy are responsible for reporting this information.

The reporting codes are:

- Yes (1) – asset has been evaluated and meets guidelines set forth in Section 2(g)(iii) of EO 13514

- No (2) – asset has been evaluated and does not meet guidelines set forth in Section 2(g)(iii) of EO 13514

- Not Yet Evaluated (3) – asset has not yet been evaluated on whether or not it meets guidelines set forth in Section 2(g)(iii) of EO 13514

- Not Applicable (4) – guidelines set forth in Section 2(g)(iii) of EO 13514 do not apply to the asset. This includes assets that will be disposed of by the end of FY 2015.

"Not Applicable (4)" is appropriate for structures and land assets. Buildings greater than 5,000 SF (465 SM) that have not had an HPSB assessment are be coded "Not yet evaluated (3)". Buildings greater than 5,000 SF (465 SM) that have had an HPSB assessment are be coded either "Yes (1) or "No (2)" using the guidance herein. Buildings 5,000 SF (465 SM) and under are not subject to the Guiding Principles, and no Federal reporting is required.

5-3.1   New Construction and Major Renovations.

During the project definition, design, and construction phases of a project, compliance with the new construction and major renovations requirements is tracked using the DoD Component checklists referenced in the paragraph entitled, "Compliance Checklists". Individual project performance is included in the Annual Energy Management Report to Congress. Refer to each DoD Component for guidance on reporting systems and respective instructions.

Once the new construction or major renovation project is complete, the status of the building is tracked in each service reporting system as detailed above. A building meeting the new construction and major renovations requirements, also meets the

---

[10] The Air Force utilizes the Automated Civil Engineer System, Real Property Module (ACES-RP), RPA Sustainability Code field.
[11] The Army utilizes the Headquarters Installation Information System (HQIIS).

Guiding Principles requirements for existing buildings and can coded as "Yes (1)". Indicating the appropriate sustainability status must be accomplished at the completion of the project when the asset record is created or updated.

5-3.2   Existing Buildings, HPSB Assessment.

Prior to reporting the status of an existing building, the building must be assessed using the requirements in Chapter 4. If the assessment verifies compliance with the Guiding Principles, code the building as "Yes (1)". If the assessment indicates that not all of the Guiding Principles have been met, code the building as "No (2) – …does not meet guidelines …". Buildings that have not yet been assessed are coded as "Not Yet Evaluated (3)".

5-4   DEPARTMENT OF DEFENSE SUSTAINABILITY POLICY REQUIREMENTS.

In accordance with OUSD AT&L Memorandum, "Department of Defense Sustainable Buildings Policy", DoD Components will design and build all new construction and major renovations projects: 1) in compliance with the Guiding Principles, 2) third-party certified to the US Green Building Council (USGBC) Leadership in Energy and Environmental Design (LEED) Silver level (or approved equivalent rating), and 3) achieve no fewer than 40% of the certification points related to energy and water conservation. In addition, all repair and renovations projects must conform to the Guiding Principles where they apply.

5-4.1   Certification Clarification of National Defense Authorization Act for Fiscal Year 2012 (Public Law 112-81) (NDAA 2012).

Section 2830(b)(1) of the NDAA 2012 prohibits the obligation or expenditure of funds authorized to be approved or otherwise made available for fiscal year 2012 for achieving any LEED Gold or Platinum certification for DoD facilities. The law allows the Secretary of Defense to waive the prohibition if a cost-benefit analysis results in a demonstrated payback and the Secretary notifies Congress at least 30 days before the obligation of funds. LEED Gold and Platinum certifications may be permitted, and not require a waiver and notification, if achieving such a certification imposes no additional cost to the Department of Defense.

5-4.2   Certification of Special Facility Types and Overseas Locations.

See individual Component policy for guidance on industrial facilities (e.g. hangars, maintenance facilities and other facilities) and non-vertical construction.

Third-party certification is required for applicable DoD component facilities in US territories and overseas on enduring installations.

Third-party certification is not required for projects in countries where the United States contributes a portion of the construction cost but does not control the design criteria (e.g. NATO funded projects). Incorporate sustainable development strategies and features to the greatest extent practical.

Third-party certification is not required for facilities being constructed by or for use by the United States under the laws, codes, rules and regulations of a Host Nation, but may need to comply with similar requirements of the Host Nation, as applicable international agreements provide. Incorporate sustainable development strategies and features to the greatest extent practical.

5-4.3   Minimum Program Requirement (MPR) 6, Sharing Whole-Building Energy and Water Usage Data.

On 25 July 2011, the U.S. Green Building Council granted a retro-active exemption from Minimum Program Requirement (MPR) 6 (Must Commit to Sharing Whole-Building Energy and Water Usage Data) for all DoD projects certified under LEED 2009 and registered for certification under LEED 2009, and to provide the exemption from MPR 6 for future projects that register for LEED certification under LEED 2009. Project teams are directed to use the "MPR 6 Department of Defense Exemption Process" (http://www.wbdg.org/pdfs/mpr6_dod_exemptprocess.pdf) for those projects certified under LEED 2009 and projects already registered for certification under LEED 2009 to retroactively request the exemption, and for future projects that will register for certification under LEED 2009.

5-4.4   Office of Secretary of Defense Guidance.

OUSD AT&L Memorandum, "Department of Defense Sustainable Buildings Policy", 25 October 2010, http://www.wbdg.org/pdfs/dod_sustainable_bldgs_policy_20101025.pdf.

OUSD AT&L Memorandum, "DoD Implementation of Storm Water Requirements under Section 438 of the Energy Independence and Security Act (EISA)", 19 January 2010, https://eko.usace.army.mil/public/fa/sdd/leed/?syspage=Documents&id=172523.

5-4.5   Air Force Guidance.

"Air Force Sustainable Design and Development (SDD) Implementation Guidance", 02 June 2011, http://www.wbdg.org/ccb/browse_cat.php?c=265.

Air Force DCS/Logistics, Installations & Mission Support Memorandum.

5-4.6   Army Guidance.

Army Supplemental Guidance to Designers and Policy Statements may be viewed at: https://mrsi.usace.army.mil/sustain/SitePages/PolicyDocuments.aspx?InitialTabId=Ribbon%2EDocument&VisibilityContext=WSSTabPersistence&.

5-4.7   Navy and Marine Corps Guidance.

NAVFAC ECB No. 2011-01, "Navy Shore Energy Building Standard", 20 December 2010, http://www.wbdg.org/ccb/browse_cat.php?c=268.

This Page Intentionally Left Blank

CHAPTER 6 SUSTAINABLE INSTALLATIONS

6-1   OVERVIEW.

Projects are not stand alone initiatives but are part of a comprehensive sustainable installation master plan that defines the holistic vision for planning and development of the installation. The value of planning is to look beyond the individual project to create synergies within the larger area, campus, neighborhood and the entire installation. The master plan provides the vision, goals and objectives for sustainable base development. It sets specific planning and site parameters that all projects must follow. It also provides documentation of the collaborative process involving stakeholders, functional advocates such as environmental, natural, historical and cultural experts, force protection officers, transportation, energy and local and regional planning managers.

For integrated design team participants, it is essential to understand that DoD installation planning has embraced the strategies of sustainable planning. Projects that comply with installation master plans may be solutions that promote compact, in-fill development and will promote mixed use (vertical or horizontal) alternatives. Project delivery teams must consider multi-story solutions that are more narrow in shape when mission requirements allow.

Project delivery teams must consider transportation connectivity and must ensure such aspects as connected sidewalks, pedestrian spaces, and bike paths are incorporated. Appropriate sustainable landscaping features must be included.

Projects must meet the requirements of the installation master plan and/or studies, including neighborhood or campus plans as applicable.

Incorporate the key principles of sustainable planning identified in UFC 2-100-01, as well as the DoD Component's individual policies and guidance, in master plans, area development plans, and other planning products..

6-2   GUIDING PRINCIPLES FOR FEDERAL LEADERSHIP IN HIGH PERFORMANCE AND SUSTAINABLE INSTALLATIONS.

Department of Defense and Military Component installation-wide policies can be used to demonstrate compliance with the Guiding Principles when there is also evidence that the policies are being implemented at the specific building. These policies may be used as models for crafting policies for building level operations.

The following list provides examples of Guiding Principles that may be met installation-wide:

- Stormwater: Strategies typically account for handling runoff associated with all buildings on a site, not just individual buildings.

- Outdoor water use reduction: In campus-type settings, outside open areas are often shared among buildings, making it difficult to assign water use to a particular building for calculating the outdoor water use reduction requirement.

- <u>Renewable energy</u>: Renewable energy generated on a contiguous campus is acceptable. Such development must be engineered to eliminate any adverse effects to electric grid frequency stability, voltage regulation, or system power factor, and will not operate in conjunction with backup generators.

- <u>Process Water</u>: Buildings served by the same central cooling equipment can claim compliance when the requirement is met.

- <u>Maximizing Efficient Use of Existing Workplaces</u>: This requirement can be considered for campus-wide space utilization..

- <u>Integrate with Local and Regional Planning</u>: A campus-wide representative could engage with local planning efforts.

- <u>Mitigate the Heat Island Effect</u>: This requirement may involve outdoor hardscapes shared by multiple buildings on a campus.

- <u>Integrated Pest Management</u>. Use integrated pest management techniques as appropriate to minimize pesticide usage. Use EPA-registered pesticides only when needed.

- <u>Environmental Tobacco Smoke Control</u>. Prohibit smoking within the building and within 25 feet (7.62 meters) of all building entrances, operable windows, and building ventilation intakes.

6-3   DOD STRATEGIC SUSTAINABILITY PERFORMANCE PLAN (SSPP).

EO 13514 requires each Federal agency to "develop, implement, and annually update an integrated Strategic Sustainability Performance Plan that will prioritize agency actions based on lifecycle return on investment (ROI)." The DoD SSPP provides an organized approach to meeting the requirements of the EO, including HPSB. Included in the DoD SSPP are the following:

- Sustainability goals, targets, schedules, milestones, and approaches

- Strategic planning and budgeting

- Activities, policies, plans, procedures, and practices

- Annual updates, opportunities for improvement, and evaluation of past performances

The DoD SSPP can be found at http://sustainability.performance.gov/.

Sustainable Communities, an Environmental Security Technology Certification Program funded project, is an on-going demonstration supporting the goals of the DoD SSPP. More information on Sustainable Communities can be found at http://www.serdp-estcp.org/Program-Areas/Resource-Conservation-and-Climate-Change/(list)/1/.

APPENDIX A REFERENCES

ADVISORY COUNCIL ON HISTORIC PRESERVATION
http://www.achp.gov/

Sustainability and Historic Federal Buildings, 02 May 2011,
http://www.achp.gov/docs/SustainabilityAndHP.pdf

AMERICAN SOCIETY OF HEATING, REFRIGERATING AND AIR-CONDITIONING ENGINEERS (ASHRAE)
http://www.ashrae.org/

ANSI/ASHRAE Standard 55-2010 (ASHRAE 55), Thermal Environmental Conditions for Human Occupancy, 2010

ANSI/ASHRAE Standard 62.1-2010 (ASHRAE 62.1), Ventilation for Acceptable Indoor Air Quality, 2010

ANSI/ASHRAE/IESNA Standard 90.1-2007 (ASHRAE 90.1), Energy Standards for Buildings Except Low Rise Residential Buildings, 2007

ANSI/ASHRAE/IESNA Standard 90.1-2010 (ASHRAE 90.1), Energy Standards for Buildings Except Low Rise Residential Buildings, 2010

ANSI/ASHRAE/USGBC/IES Standard 189.1-2011 (ASHRAE 189.1), Standard for the Design of High-Performance Green Buildings Except Low-Rise Residential Buildings, 2011

ENVIRONMENTAL PROTECTION AGENCY

EPA 841-B-09-001, Technical Guidance on Implementing the Stormwater Runoff Requirements for Federal Projects under Section 438 of the Energy Independence and Security Act, December 2009, http://www.epa.gov/owow/NPS/lid/section438/

Clean Air Act of 1990 as amended in 1997, http://epw.senate.gov/envlaws/cleanair.pdf

Clean Water Act of 1977, as amended, http://epw.senate.gov/water.pdf

Comprehensive Procurement Guideline (CPG) Program, http://www.epa.gov/epawaste/conserve/tools/cpg/index.htm

Endangered Species Act of 1973, as amended, http://epw.senate.gov/esa73.pdf

Energy Independence and Security Act (EISA), 19 December 2007, http://www.gpo.gov/fdsys/pkg/BILLS-110hr6enr/pdf/BILLS-110hr6enr.pdf

Energy Policy Act of 1992 (EPAct 1992), as amended, http://thomas.loc.gov/cgi-bin/query/z?c102:H.R.776.ENR:

Energy Policy Act of 2005 (EPAct 2005), http://www.gpo.gov/fdsys/pkg/PLAW-109publ58/pdf/PLAW-109publ58.pdf

ENERGY STAR® Portfolio Manager, http://www.energystar.gov/index.cfm?c=evaluate_performance.bus_portfoliomanager

Resource Conservation and Recovery Act (RCRA), 21 October 1976, http://www.epa.gov/agriculture/lrca.html

WaterSense Program, http://www.epa.gov/watersense

GENERAL SERVICES ADMINISTRATION

Green Products Compilation (GPC), http://www.gsa.gov/greenproductscompilation

INSTITUTE OF ELECTRICAL AND ELECTRONICS ENGINEERS

IEEE 1547, Standard for Interconnecting Distributed Resources with Electric Power Systems, 2003, http://grouper.ieee.org/groups/scc21/1547/1547_index.html

INTERNATIONAL INSTITUTE FOR SUSTAINABLE LABORATORIES

Labs21® Energy Benchmarking Tool, http://www.i2sl.org/resources/toolkit/benchmark.html

INTERNATIONAL ASSOCIATION OF PLUMBING AND MECHANICAL OFFICIALS
http://www.iapmo.org/

Uniform Plumbing Code, 2012

INTERNATIONAL CODE COUNCIL
http://www.iccsafe.org/

International Plumbing Code, 2012

NATIONAL INSITUTE OF BUILDING SCIENCES, WHOLE BUILDING DESIGN GUIDE
http://www.wbdg.org/

Federal Green Construction Guide for Specifiers, http://www.wbdg.org/design/greenspec.php

NATIONAL INSTITUTE OF STANDARDS AND TECHNOLOGY

NIST Handbook 135, Life-Cycle Costing Manual for the Federal Energy Management Program, http://fire.nist.gov/bfrlpubs/build96/PDF/b96121.pdf

SHEET METAL AND AIR CONDITIONING CONTRACTOR'S NATIONAL ASSOCIATION (SMACNA)
www.smacna.org/

SMACNA 008, *Indoor Air Quality (IAQ) Guidelines for Occupied Buildings Under Construction, 2nd Edition*, 2007

UNITED STATES CODE
http://www.gpo.gov/fdsys/browse/collectionUScode.action?collectionCode=USCODE

USC Title 7 Section 4201, FARMLAND PROTECTION POLICY – General Provisions

USC Title 10 Section 2853, ADMINISTRATION OF MILITARY CONSTRUCTION AND MILITARY FAMILY HOUSING – Authorized Cost and Scope of Work Variations

USC Title 16 Section 470, National Historic Preservation Act

USC Title 42 Chapter 55, NATIONAL ENVIRONMENTAL POLICY – National Environmental Policy Act of 1969

UNITED STATES CODE OF FEDERAL REGULATIONS
http://www.gpo.gov/fdsys/browse/collectionCfr.action?collectionCode=CFR

CFR Title 10 Part 433, Energy Efficiency Standards for New Federal Commercial and Multi-Family High-Rise Residential Buildings

CFR Title 10 Part 434, Energy Code for New Federal Commercial and Multi-Family High-Rise Residential Buildings

CFR Title 10 Part 435, Energy Efficiency Standards for New Federal Low-Rise Residential Buildings

CFR Title 10 Part 436, Federal Energy Management And Planning Programs

CFR Title 36 Part 800, Protection of Historic Properties

UNITED STATES DEPARTMENT OF AGRICULTURE

BioPreferred Program, http://www.biopreferred.gov/

UNITED STATES DEPARTMENT OF DEFENSE

DoD 4160.21-M, Defense Materiel Disposition Manual, 28 May 2004

DoDI 4170.11, Installation Energy Management, 11 December 2009,
http://www.dtic.mil/whs/directives/corres/ins1.html

DoDI 4715.16, Cultural Resources Management, 18 September 2008,
http://www.dtic.mil/whs/directives/corres/ins1.html

National Defense Authorization Act (NDAA) of 2012, 31 December 2011,
http://www.gpo.gov/fdsys/pkg/BILLS-112hr1540enr

UNITED STATES DEPARTMENT OF DEFENSE, LEGACY RESOURCE MANAGEMENT PROGRAM
http://www.denix.osd.mil/cr/LRMP/FactSheets.cfm

Legacy Project Number 04-220, DoD Sustainability Application Guide for Historic Properties, May 2007, http://www.denix.osd.mil/cr/upload/04-220_FINAL_0.PDF

Legacy Project Number 09-451, A Case Study for Preserving a Department of Defense Historic Building and Achieving LEED Certification for a Major Renovation Project Indiana Army National Guard Stout Field, Building 5, March 2011

Legacy Project Number 09-452, Strategies for Greening Historic Properties, August 2010

Legacy Project Number 09-456, Maintaining Elements that are Efficient by Design: What's Already Green about Our Historic Buildings, July 2010

UNITED STATES DEPARTMENT OF DEFENSE, UNIFIED FACILITIES CRITERIA PROGRAM
http://dod.wbdg.org/

UFC 1-200-01, General Building Requirements

UFC 2-100-01, Installation Master Planning

UFC 3-101-01, Architecture

UFC 3-110-03, Roofing

UFC 3-201-02, Landscape Architecture

UFC 3-210-10, Low Impact Development

UFC 3-400-02, Design: Engineering Weather Data

UFC 3-530-01, Design: Interior and Exterior Lighting and Controls

UFC 3-600-01, Fire Protection Engineering for Facilities

UFC 3-701-01, DoD Facilities Pricing Guide

UNITED STATES EXECUTIVE ORDERS
http://www.archives.gov/federal-register/executive-orders/

Executive Order 13006, Locating Federal Facilities on Historic Properties in Our Nation's Central Cities, 21 May 1996.

Executive Order 13150, Federal Workforce Transportation, 21 April 2000.

Executive Order 13221, Energy-Efficient Standby Power Devices, 31 July 2001.

Executive Order 13327, Federal Real Property Asset Management, 4 February 2003.

Executive Order 13423, Strengthening Federal Environmental, Energy, and Transportation Management, 27 January 2007.

Executive Order 13514, Federal Leadership in Environmental, Energy, and Economic Performance, 5 October 2009.

Guiding Principles for Federal Leadership in High Performance and Sustainable Buildings (Guiding Principles), "High Performance and Sustainable Buildings Guidance" (Updated December 2008), http://www.wbdg.org/references/fhpsb.php

UNITED STATES NATIONAL PARK SERVICE

National Park Service, Improving Energy Efficiency in Historic Buildings, Preservation Brief 3, http://www.nps.gov/hps/tps/briefs/brief03.htm

Secretary of Interior, Standards for Rehabilitation, Illustrated Guidelines for Rehabilitating Historic Buildings, http://www.nps.gov/tps/standards/rehabilitation/rehab/index.htm

Secretary of Interior, Guidelines for the Treatment of Cultural Landscapes, http://www.nps.gov/tps/standards/four-treatments/landscape-guidelines/index.htm

This Page Intentionally Left Blank

APPENDIX B BEST PRACTICES

B-1   MOISTURE CONTROL PLANNING.

The Moisture Control Plan is an organized plan for good design practice. The prevention of mold and moisture problems is always part of good design detailing and specifications.

The Moisture Control Plan must be written to follow http://www.wbdg.org/resources/moisturedynamics.php?r=fhpsb_new and the construction documents must comply with each paragraph under "D. Design Recommendations", "E. Methods of Construction", and one of the two methods in "F. Analysis and Design Tools."

At a minimum, the Moisture Control Plan must also address:

- Moisture control through design details for the envelope and for HVAC humidity controls

- Specified plan for construction materials storage and protection from mold and moisture infiltration

- Specified humidity control during construction

- Operations and Maintenance plan (to include warranty and follow-on maintenance contracts) for building maintenance to control moisture after Beneficial Occupancy Date.

B-2   LIGHTING.

Lighting: Design interior and exterior lighting systems in accordance with UFC 3-530-01. Design lighting systems to maximize energy savings, reduce maintenance costs, and improve lighting quality, utilizing energy efficient technologies and controls. Lighting systems must be designed; simple retrofits will only yield minimum benefit.

Exterior Lighting: Design exterior lighting systems to improve the night time visibility. Illuminate areas to the appropriate levels, improve uniformity, and reduce light trespass, light pollution and glare. Maximize energy savings.

Interior Lighting: Design lighting systems to improve lighting quality by uniformity, illuminating surfaces, and reducing glare. Design interior lighting systems to reduce ambient lighting to effective illuminance levels and illuminate the task. Adjustments to individual task needs should be accomplished with task lighting in lieu of ambient lighting. Provide controls for lighting systems in response to occupancy and daylighting.

B-3   RENEWABLE SYSTEMS AND ENERGY SECURITY.

Designs should address the following issues concerning the utilization of renewable energy (RE) systems:

B-3.1 Low Energy Security Risk RE Systems.

Waste-to-energy, geothermal, or biomass, are low energy security risk RE systems and are highly desirable.

- No power quality issues
- Stable, reliable, easily controllable
- Easily integrated into microgrid topology
- Straight forward, easily implemented IEEE 1547 standard

B-3.2 High Energy Security Risk RE Systems.

Photovoltaic and wind renewable energy (RE) systems are high energy security risk due to their adverse effect on electric grid frequency control, voltage regulation, and power quality.

B-3.2.1 Wind Generators.

- Consume reactive power – cannot provide reactive power to meet needs of base
- Lower utility power factor (PF) during operation – creates risk of utility power factor penalty charges or power outages
- Electromagnetic interference
- Implementation of IEEE 1547 standard is more complex

B-3.2.2 PV Systems.

- Provide near unity PF – little to no reactive power
- Lower utility power factor during operation
- Unstable effect on system frequency control
- Additional stress on emergency generators supporting critical facilities if operated in parallel
- Implementation of IEEE 1547 standard is more complex

B-3.3 Domestic Hot Water Solar Heating.

Implementation of solar heating for domestic hot water in the 60 to 85 percent range may be economical. Daylighting units, solar air heating system, and transpired – air solar systems are may be economical in areas with more than 1,200 equivalent full sun hours per year.

APPENDIX C GLOSSARY

ACRONYMS

ACHP	Advisory Council on Historic Preservation
AFCEE	Air Force Center for Engineering and the Environment
AFCEC	Air Force Civil Engineer Center
AFARS	Army Federal Acquisition Regulation Supplement
AFFARS	Air Force Federal Acquisition Regulation Supplement
AHJ	Authority Having Jurisdiction
APF	Appropriated Funding
ASHRAE	American Society of Heating, Refrigerating and Air Conditioning Engineers
ASTM	American Society of Testing and Materials
BEAP	Base Exterior Architecture Plan
BEI	Business Enterprise Integration (OSD Office)
CCB	Construction Criteria Base
cf	Cubic Feet
CFR	Code of Federal Regulations
CONUS	Continental United States
DFARS	Defense Federal Acquisition Regulation Supplement
DOE	Department of Energy
DOD	Department of Defense
DODI	Department of Defense Instruction
ECA	Energy Compliance Analysis
EEIM	Enterprise Energy Information Management
EISA	Energy Independence and Security Act
EMS	Environmental Management System
EO	Executive Order

EPA	Environmental Protection Agency
EPAct	Energy Policy Act
EPEAT	Electronic Product Environmental Assessment Tool
FAR	Federal Acquisition Regulation
FEMP	Federal Energy Management Program
GBCI	Green Building Certification Institute
GBI	Green Building Institute
GPF	Gallons Per Flush
GSA	General Services Administration
HP	Horsepower
HPSB	High Performance Sustainable Building
HQUSACE	Headquarters, U.S. Army Corps of Engineers
HVAC	Heating, Ventilation and Air Conditioning
IAQ	Indoor Air Quality
IBC	International Building Code
ICC	International Code Council
IESNA	Illuminating Engineering Society of North America
IEQ	Indoor Environmental Quality
KWh	Kilowatt Hour
KW	Kilowatt
LCA	Life Cycle Assessment
LCC	Life Cycle Cost
LCCA	Life Cycle Cost Analysis
LEED	Leadership in Energy and Environmental Design
LID	Low Impact Development
LPF	Liters Per Flush

M & V	Measurement and Verification
MCWB	Mean Coincident Wet-Bulb
MILCON	Military Construction
MOU	Memorandum of Understanding
NAF	Non-Appropriated Funding
NAVFAC	Naval Facilities Engineering Command
NIST	National Institute of Standards and Technology
NMCARS	Navy Marine Corps Acquisition Regulation Supplement
OCONUS	Outside Continental United States
ODS	Ozone Depleting Substances
OMB	Office of Management and Budget
PL	Public Law
PRV	Plant Replacement Value
RCRA	Resource Conservation and Recovery Act
RPA	Real Property Accountability
SHPO	State Historic Preservation Office
SMACNA	Sheet Metal and Air Conditioning Contractors' National Association
SNAP	Significant New Alternatives Policy
SPiRiT	Sustainable Project Rating Tool
UFC	Unified Facilities Criteria
UFGS	Unified Facilities Guide Specifications
USGBC	U.S. Green Building Council
US	United States
USACE	U.S. Army Corps of Engineers
USC	U.S. Code
USDA	United States Department of Agriculture

WBDG   Whole Building Design Guide

DEFINITION OF TERMS

<u>**Acquisition:**</u> A multi-disciplinary process encompassing a variety of functions that involve all facets of life cycle management including determination of need, planning; design; construction, lease, or purchase; sustainment, modernization and disposal of military installations and facilities as well as other goods and services. Acquiring supplies and services is a process governed by Federal, Defense, and service branch specific Acquisition Regulations (FAR, DFARS, and AFFARS, AFARS, or NMCARS) and Public Law (PL). Together there are more than 4,000 documents controlling the acquisition process. The acquisition process depends upon: (1) who the customer is, (2) the source of the item to be acquired, and (3) what type of funds are to be used, Appropriated (APF) or Non-Appropriated (NAF). In general, the Military Construction Program process is funded by Appropriated Funds. However, projects for Non-Appropriated Funds activities (commissaries, bowling centers, etc.) use NAF funds for building construction, upgrades, special equipment, and furnishings. APF funding is used most typically for demolition and environmental work on NAF activity sites. A combination of APF and NAF funds may be used sometimes.

<u>**Authority Having Jurisdiction (AHJ):**</u> The term "Authority Having Jurisdiction" (AHJ) as used in the codes and standards referenced in this UFC mean the component office of responsibility, i.e., U.S. Army, HQ USACE/CECW-CE; U.S. Navy, NAVFACENGCOM HQ Code CHE; U.S. Marine Corps, HQMC Code LFF-1; and U.S. Air Force, AFCEC. The enforcement of the codes and standards as they pertain to facility projects can be delegated to the local Component Office's Chief Engineer's Technical Representative at the discretion of the component's aforementioned office.

<u>**Building Automation System (BAS):**</u> An energy management system, usually with additional capabilities, relating to the overall operation of the building in which it is installed, such as equipment monitoring, protection of equipment against power failure, building security, etc.

<u>**Charrette:**</u> An intensive creative work session in which a design team focuses on a particular design problem and arrives at a collaborative solution with stakeholders from the project area. A charrette can be a breakthrough event that helps creates a meaningful master plan or facility design. Properly executed, this technique can produce a master plan or facility design that is more useful, better understood, and more quickly produced than one formed by any other method.

From the Public Involvement Techniques, publication no. FHWA-PD-96-031 HEP-30/9-96/(4M)QE

<u>**Commissioning:**</u> (This process is not funded as a program line item.) Ensures that a facility and its components will perform as designed and intended.

Optimally the commissioning process starts during planning and programming prior to design. It continues during design with review of design information, performance data in specifications, and evaluation of submissions. During construction the commissioning process includes setup of pre-construction cause and effect schedules, tests and test data, and establishes record documentation requirements.

Commissioning includes witnessing field tests, adherence to specified performance criteria such as cleaning of ducts, performance of equipment, and training of facility maintenance personnel. Commissioning of building systems is critical to ensuring their expected operation.

DD Form 1391: A programming document used by the Department of Defense to submit requirements and justifications in support of funding requests for military construction to Congress.

Energy Compliance Analysis (ECA): The ECA must identify specific energy conservation criteria that applies to the project, the software used to prepare the necessary calculations, a summary of all input to and output from the calculations, and the calculated energy consumption of the proposed design.

Energy Modeling: The process by which conceptual designs, including size, material choices, factors such as site, solar, and wind orientations, daylighting percentages, and energy system choices (solar water heat, underfloor vs. overhead air distribution systems) are analyzed to show how to optimize these factors for efficient building operation and resource consumption.

Environmental Management System (EMS): A framework that helps to achieve environmental goals through consistent control of operations. The assumption is that this increased control will improve environmental performance.

Energy Management Control Systems (EMCS): A computer/processor-based hardware and software system with sensors, control devices, and all the necessary components that perform some or all of the following functions: (a) Measures conditions related to the use of various forms of energy by HVAC systems; (b) Controls these conditions at selected set-points; (c) Monitors and/or controls the energy use; (d) Provides status reports on the HVAC system's performances; (e) Provides information for the management of a building's environment, and/or its energy efficiency, and/or HVAC system maintenance.

Environmentally Preferable Products: Products or services having a lesser or reduced effect on human health and the environment when compared with competing products or services serving the same purpose. This comparison may consider raw materials acquisition, production, manufacturing, packaging, distribution, reuse, operation, maintenance, or product or service disposal.

Green Procurement: The DoD Green Procurement Policy requires the services to have preferential purchasing programs for EPA recycled content products designated in the Comprehensive Procurement Guideline (CPG) Program, USDA biobased products; DOE Energy Star products; FEMP Energy Efficient Standby Power Devices; DOE Water Conserving products; California Low VOV products; EPA Asbestos Alternative products; products that reduce EPA priority chemicals (cadmium, lead, PCB, mercury and naphthalene); and any other products that are environmentally preferable according to EPA criteria. These products must be included in all projects whether accomplished by contract or in-house. The *USAF Guide to Green Purchasing* at www.afcee.lackland.af.mil/eq/ap/gg/guide.doc is an Internet-based resource that provides additional details.

Historic Properties: Properties that are included in the National Register of Historic Places or that meet the criteria for the National Register.

Indoor Environmental Quality (IEQ): The condition or state of the indoor built environment. Aspects of IEQ include: light quality, acoustic quality, and air quality.

Life-Cycle Costing (LCC): An important economic analysis used in the selection of alternatives that impact both pending and future costs. It compares initial investment options and identifies the least cost alternatives for a 40-year period. As applied to building design energy conservations measures, the process is mandated by law and is defined in 10 CFR Part 426, Subpart A: Program Rules of the Federal Energy Management Program (NIST Handbook 135). The National Institute of Standards and Technology has established the Building Life-Cycle Cost (BLCC) computer program to perform LCC analyses. The program incorporates user entered data for and compares the following: Sunk Costs, First Costs, Salvage Value, Future Investment, Residual Value, Annually Recurring Fixed Costs, Annually Recurring Escalating Costs, and Energy (Fuel Costs) Escalation Rates.

Life Cycle Cost Analysis (LCCA): A method for assessing the total cost of facility ownership. It takes into account all costs of acquiring, owning, and disposing of a building or building system. LCCA is especially useful when project alternatives that fulfill the same performance requirements, but differ with respect to initial costs and operating costs, have to be compared in order to select the one that maximizes net savings.

Light Pollution: The spread of artificial light which is not used for its desired function and is wasted into the night sky. Such wasted light has very detrimental effects both for energy savings and for the native plants and animals which use the site and its surroundings."

Planning: Initiated by a using entity whose facility needs are driven by mission requirements. The insertion of facility projects into the official budget is based on the requirement for mission support. As mission priorities change, projects can be moved ahead of their initial scheduled fiscal year or alternately, be dropped altogether.

PRV (Plant Replacement Value): PRV is the cost of replacing the current physical plant with modern facilities built at today's construction costs using today's construction standards. It includes all buildings, structures, or other improvements to real property, regardless of funding source but does not include land value or leasing costs.

Programming: The data collection process done by the military project management team in order to achieve a conceptual design in enough detail to prepare a viable cost estimate. Programming data is reported on DD Form 1391 for design-bid-build and design-build projects, and is scrutinized from Installation-level through the Office of

Management and Budget (OMB), Congress, and signed by the President before it is approved as a project.

Project Definition: As the military programming process often predates actual design by two to five years or more, when the design services portion of a project is funded, the project design team must validate stated project requirements and personnel assignments based upon updated mission criteria. This process is sometimes referred to as Project Definition to distinguish it from the usual private sector A/E reference to "programming".

Project Scope: Refers to the physical size, cost and mission of a project. With historical data on facilities for various missions, experienced programmers will formulate the conceptual design for the DD Form 1391 based upon the required size of a facility to accommodate mission functions. Once the scope for a project has been enacted by Congress, changing the scope of a project will bring delay a project until the revision has been justified and approved. This is true whether or not the costs have also changed. Reductions of project scope (by no more than 25%) will require Congressional notification, review, and authorization before proceeding. Increasing the project scope is not authorized without reprogramming. See USC Title 10 Section 2853 for more information.

Recommissioning: A systematic process for investigating, analyzing and optimizing the performance of a building through operational and maintenance improvement measures and ensuring their continued performance over time.

Retrocommissioning: Applies a quality assurance (commissioning) process "retroactively" to an existing building.

Total Ownership Cost (TOC): Total of all direct and indirect costs associated with an asset or acquisition over its entire life cycle.

View Shed: An area of land, water, or other environmental element that is visible to the human eye from a fixed vantage point.

Xeriscaping: Quality landscaping that conserves (reduces or eliminates) the need for supplemental irrigation water and protects the environment.

This Page Intentionally Left Blank

APPENDIX D PROJECTS IMPACTING A HISTORIC BUILDING OR DISTRICT

D-1 APPLICABILITY TO SUSTAINABILITY GOALS.

Most buildings listed in or eligible for listing in the National Register of Historic Places are capable of meeting the Guiding Principles and should work toward compliance with energy goals. Following the National Historic Preservation Act of 1966 (NHPA) Section 106 implementing regulations, CFR Title 36 Part 800, will facilitate the consultation needed to meet sustainability goals in proposed construction/rehabilitation/repair projects. This process will also take into account indirect impacts that the project could have on other cultural resources that could trigger project delays. Early inclusion of the DoD Component -specific Installation Cultural Resource Manager who meets the Secretary of Interior Professional Standards is the best deterrent to later project delays relative to these issues.

Sustainably rehabilitating historic buildings conserves the invested-cost of energy and materials as well as encouraging the preservation of these resources and cultural landscapes. The design, materials, type of construction, size, shape, site orientation, surrounding landscape and climate all play a role in how buildings perform. Before implementing any energy conservation measures to enhance the sustainability of a historic building, the existing energy-efficient characteristics of the building should be assessed as historic construction methods and materials often maximized natural sources of heating, lighting and ventilation to respond to local climate conditions.

Some projects may impact directly or indirectly other cultural resources classified as historic properties. For example, new construction could be located on a site with archaeological resources, which are also covered by the NHPA. Early inclusion of the component-specific cultural resource professional will ensure all cultural resources are assessed.

For DoD's historic facilities, certain generic sustainable upgrades may not be the most appropriate solution. For such buildings, it is vital to involve trained preservation professionals and the State Historic Preservation Office (SHPO) early in the design process to achieve maximum success in the process while also taking into consideration the character-defining features of the buildings during the design process. Historic buildings represent a previous long-term investment of resources and energy and as such these buildings were often constructed using methods and materials that maximized natural sources of heating, lighting, and ventilation to respond to local climatic conditions. Work with the Installation Cultural Resource Manager and SHPO to identify these elements of historic buildings that are functional passive design components, or inherently sustainable in nature. Useful tools may also include the Secretary of Interior's Guidelines for Rehabilitation on Sustainability for Rehabilitating Historic Buildings, the National Park Service's *Improving Energy Efficiency in Historic Buildings*, Preservation Brief 3 and the Advisory Council on Historic Preservation's *Sustainability and Historic Federal Buildings*.

D-2   NEW CONSTRUCTION AND MAJOR RENOVATION PROJECTS.

Major renovation of historic facilities and new construction near historic facilities must follow the Secretary of Interior's *Standards for Rehabilitation, Illustrated Guidelines for Rehabilitating Historic Buildings* with the aim of avoiding the creation of an adverse effect requiring mitigation. The Installation will be responsible for ensuring compliance with the NHPA for these renovation/repair projects. Coordinate all projects involving historic properties with the Installation Cultural Resource Manager who can facilitate consultation with the SHPO prior to design.

D-3   REPAIR AND MAINTENANCE PROJECTS.

Repair and maintenance to existing buildings must follow the Secretary of Interior's *Standards for Rehabilitation, Illustrated Guidelines for Rehabilitating Historic Buildings* with the aim of avoiding the creation of an adverse effect requiring mitigation. The Installation will be responsible for ensuring compliance with the NHPA for these renovation/repair projects. Coordinate all projects involving historic properties with the SHPO via the Installation Cultural Resource Manager prior to design. Information pertaining to historic buildings and districts is available in the Installation Cultural Resources Management Plan, as required by DoDI 4715.16.

D-4   LEASED HISTORIC BUILDINGS.

Responsibilities for historic facilities that are also leased buildings must be in the terms of the lease. Renovations or repairs must follow the Secretary of Interior's *Standards for Rehabilitation, Illustrated Guidelines for Rehabilitating Historic Buildings* to avoid creating an adverse effect. Allow enough time for the proper parties including the DoD Component Installation Cultural Resource Manager, identified in the lease, to complete consultation with the SHPO and other stakeholders.

D-5   NHPA SECTION 106 PROCESS.

Section 106 of the NHPA requires Federal agencies to take into account the effects of their undertakings on historic properties, and afford the Advisory Council on Historic Preservation a reasonable opportunity to comment. Process description is available at http://www.achp.gov/106summary.html.

Some projects may impact directly or indirectly other cultural resources classified as historic properties. For example, new construction could be located on a site with archaeological resources, which are also covered by the NHPA. Early inclusion of the component-specific cultural resource professional will ensure all cultural resources are assessed.

D-5.1   Initiate Process.

Determine whether or not the project involves a type of activity that could affect historic properties. If so, identify the appropriate SHPO to consult with during the process. Identify other potential consulting parties. If there are no activities that have potential to affect historic properties, no Section 106 obligations exist.

D-5.2 Identify Historic Properties.

Review background information, consult with the SHPO and others, seek information from knowledgeable parties, and conduct additional studies as necessary. Districts, sites, buildings, structures, and objects listed in the National Register are considered; unlisted properties are evaluated against the National Park Service's published criteria, in consultation with the SHPO and any Indian tribe or Native Hawaiian organization that may attach religious or cultural importance to them. If questions arise about the eligibility of a given property, the agency may seek a formal determination of eligibility from the National Park Service.

If there are no historic properties present or affected provide documentation to the SHPO and, barring any objection in 30-days, proceed with the project. If there are historic properties present, proceed to assess possible adverse effects.

D-5.3 Assess Adverse Effects.

In consultation with the SHPO, make an assessment of adverse effects on the identified historic properties based on criteria found in Advisory Council on Historic Preservation's (ACHP) regulations.

If they agree that there will be no adverse effect, proceed with the project and any agreed-upon conditions.

If they find that there is an adverse effect, or if the parties cannot agree and ACHP determines within 15-days that there is an adverse effect, begin consultation to seek ways to avoid, minimize, or mitigate the adverse effects.

D-5.4 Resolve Adverse Effects.

Consult to resolve adverse effects with the SHPO and others, who may include Indian tribes and Native Hawaiian organizations, local governments, permit or license applicants, and members of the public. The ACHP may participate in consultation when there are substantial impacts to important historic properties, when a case presents important questions of policy or interpretation, when there is a potential for procedural problems, or when there are issues of concern to Indian tribes or Native Hawaiian organizations.

Consultation usually results in a Memorandum of Agreement (MOA), which outlines agreed-upon measures that the agency will take to avoid, minimize, or mitigate the adverse effects. In some cases, the consulting parties may agree that no such measures are possible, but that the adverse effects must be accepted in the public interest.

D-5.5 Implementation.

If an MOA is executed, proceed with the undertaking under the terms of the MOA.

D-5.6 Failure to Resolve Adverse Effects.

If consultation proves unproductive, the Agency or the SHPO, or ACHP itself, may terminate consultation. If a SHPO terminates consultation, the Agency and ACHP may conclude an MOA without SHPO involvement. Submit appropriate documentation to ACHP and request ACHP's written comments. Take into account ACHP's written comments in deciding how to proceed.

D-5.7 Tribes, Native Hawaiians, and the Public.

Major emphasis is placed on consultation with Indian tribes and Native Hawaiian organizations, in keeping with the 1992 amendments to NHPA. Consultation with an Indian tribe must respect tribal sovereignty and the government-to-government relationship between the Federal Government and Indian tribes. Even if an Indian tribe has not been certified by the National Park Service (NPS) to have a Tribal Historic Preservation Officer who can act for the SHPO on its lands, it must be consulted about undertakings on or affecting its lands on the same basis and in addition to the SHPO.

D-6 SPECIAL CONSIDERATIONS.

D-6.1 Daylighting.

If windows, skylights and other elements allowing for daylighting are character defining features on the historic building, discuss options with the Installation Cultural Resource Manager.

If daylighting principles do not cause an adverse effect to the historic facilities, then employ to the fullest extent practicable.

D-6.2 Solar Panel Installation.

The following website provides advice on projects from solar panel installation to heating and air conditioning upgrades to weatherization and insulation in a sensitive and cost effective manner: http://www.nrel.gov/docs/fy11osti/51297.pdf

D-6.3 Infill.

Infill is appropriate within historic districts or neighborhoods if carried out in consultation with all stakeholders and sensitively executed. Spatial information on installation historic resources is available in the Integrated Cultural Resources Management Plan as required by DoDI 4715.16.

D-6.4 Cistern Use.

The use of cisterns, either sub-grade or above-ground, were a frequent mechanism for water storage in many historic buildings and structures and may be an appropriate addition when undertaking a major building renovation.

If a historic cistern is available at the site, reuse the location and equipment, as practicable.

D-6.5   Replacement of Materials.

Rehabilitation of historical buildings must allow in-kind replacement of materials where an adverse effect will result if not done.

D-6.6   Reclaimed/Salvaged Materials.

When repairing or renovating historic buildings, reclamation of salvaged materials must be used instead of procurement or repair to obtain assets to meet current requirements, when timely and economical to do so (DoDI 4715.16 Paragraph 6.1). Monetary value may not be considered a limitation if reclamation will satisfy critical item requirements or if new procurement/production is impractical. The DoD Component Installation Cultural Resource Manager can assist in this assessment. See DoD 4160.21-M Chapter 13 (Reclamation) for more information.

D-7   LIFE CYCLE COST ANALYSIS CONSIDERATIONS.

Ensure costs related to the mitigation of adverse effects to historical properties are included in the Life Cycle Cost Analysis (LCCA).

D-8   HISTORICAL LANDSCAPES.

Consider historic landscapes when planning water efficient landscapes and irrigation systems. See Secretary of Interior's *Guidelines for the Treatment of Cultural Landscapes* for additional information.